STAIRWAY TO HELL

Discussion Guide

Includes Story, Q&A and Facilitator's Tips

ENERGYPHILE SESSION №1

This story incorporates fictional elements inspired by historical artifacts and is based on the author's own research. Certain names, characterizations, descriptions and scenarios are either products of the author's imagination or used in a fictitious manner.

Where deemed necessary, every reasonable effort has been made to contact the copyright holders for work reproduced in this book.

Published by Energyphile Media Inc.
energyphile.org

ISBN 978-1-9991113-4-2 (paperback)
ISBN 978-1-9991113-5-9(ebook)

Produced by Page Two
pagetwo.com

Edited by Lori Burwash
Cover and interior design by Taysia Louie
Original design concept by Christina Sweetman

Contents

PREAMBLE

Popular business culture is awash with success stories. Tech entrepreneurs who go from zero to billions are feted in books and magazines and across every media channel. Their accomplishments are impressive and often inspiring.

But what about failure stories? We don't hear them quite as often. That's because demise is a lot harder to acknowledge and discuss. But, as you and I both know, much can be learned from failure. Some argue that falling down a few times is a prerequisite for success. I tend to agree.

"Stairway to Hell" is about corporate failure in the extreme. So extreme that the 1922 insolvency of a local coal mining company wiped out Bankhead, a once-vibrant community of 1,500 people in the Canadian Rockies.

I've been to Bankhead at least 10 times, wandering through its deserted remains. As I've explored the ghost town, I've relived the spirit of its people, its history and its tragedies — and I've learned a lot from its downfall. What's eerie is how much this story can be juxtaposed with corporate tales from the energy business today.

While Mine No. 80 was plagued by many issues, including low-quality product and high-cost production, "Stairway to Hell" boils down to one tenet: hope is not a strategy. Cliché, yes, but, in the slipstream of failing enterprises, it's a pervasive dictum.

Strangely, the tale of Bankhead's demise is more inspiring to me than a typical success story. Its insights into what causes failure can be flipped around to reveal lessons for success for us all.

“**What the hell**
happened here?”

STAIRWAY TO HELL

JASON AND KATIE are in love. It says so on the crumbling concrete wall in front of me. Or at least they were in love, a long time ago, when Jason took a pencil and firmly inscribed his name above Katie's inside a big heart.

But this story isn't about love. It's about economic misery. It's about how quickly an entire town of workers, families, children and young lovers can descend into a pit of nothingness. It's about decrepit homes, factories filled with rusting equipment and what's left of a church on a hill.

No Signs of Life

I'm walking around Bankhead, at the base of Cascade Mountain in the Canadian Rockies. Hidden behind trees in a picturesque valley, Bankhead is an old coal mining village at latitude 51.231550 and longitude -115.522143 that was once home to

1,500 people. It's good to know the coordinates because there are no streets left in Bankhead. The only sign that this ghost town exists is a small brown one on a secondary Alberta highway.

However, Bankhead isn't completely lost to the world. Most days, a few curious sightseers come from the nearby tourist beehive of Banff. On the road to Lake Minnewanka, they catch sight of that brown sign, turn into a parking lot and maybe have lunch at one of the lonely-looking picnic tables. From there, a short hiking trail loops a kilometer or two around what used to be the town's mining operation. Plaques on squat guideposts inform visitors of the local history and point out what remains of the landmarks on the multi-acre site.

I'm standing in one of Bankhead's empty buildings. The dirt floor is littered with pine cones, one of which I inadvertently step on. I look up and get the impression the sky doesn't like what I just did. Bad-tempered clouds spit drops of rain on me through the long-gone roof. But I'm not bothered. This isn't a happy place. In fact, bad weather seems apropos to the mood around here.

Across the room, a branch pokes through an empty window frame. I wonder, Which had less pity for this economic tragedy: business or nature?

There isn't much doubt in my mind. I know some of the history around here. Business was brutal in 1922, the year everything crashed to an end for Bankhead. Back then, cigar-smoking tycoons packed more punch than anything nature could throw at this town. A boardroom decision to shutter the mine was more devastating than any winter storm passing through the place. Looking around at Bankhead's ghostly remnants, I come to an uncomfortable conclusion: resource-based communities live under the constant threat of oblivion.

The lamp house was once the gateway to the economic heart of Bankhead. Now, the crumbling walls that remain are covered with graffiti.

History's Slag Heap

As someone who studies the energy business, I'm alert to large-scale obsolescence. More than 85% of the world's insatiable needs are extracted from the earth's crust in the form of coal, oil, natural gas and uranium — and plenty of people on six out of seven continents make a living bringing that stuff out of the ground. Now the pressures of change threaten many resource-dependent towns, regions, even entire countries. Their future viability is always hostage to environmental pressures as well as to the dragon that eventually burns every industry in the ass: technological change.

My research into energy transitions, along with field trips to places like Bankhead, has taught me that relegating a business to history's slag heap can happen faster than a bucket of water can soak a lump of burning coal. One day everyone has a job, the next a thousand people are walking the streets in a daze with nothing to do. Years later, mischievous young lovers scrawl graffiti on what used to be the heart of a community's economic well-being, and inside of a century, a curious scribe like me wanders through the rubble asking, "What the hell happened here?"

Do the remains of Bankhead presage the fate of today's oil-producing centers?

In this new age of renewable energy, the parallels are easy to draw. Migration to sources like wind and solar on the front end of energy supply — combined with new alternative devices like electric cars on the consumption end — potentially threatens the livelihoods of millions of people around the world involved in the production of earthly fuels.

It's not just me pretending to be Socrates, philosophizing on a dreary, ash-colored day. Entire countries are thinking about the threat of irrelevance — places like Saudi Arabia, the world's second-largest oil producer. In 2014, policy makers in the desert kingdom had an ontological moment. Recognizing excessive dependency on their potentially limited fuel source, they began thinking about how to diversify away from their oil-drenched economy.

This old mining town is an acute case study that should make **anyone think twice about being complacent about their lot in life.**

Okay, compared to Saudi Arabia, Bankhead looks insignificant, a kingdom only in the minds of those who once lived where I now stand. But scale doesn't matter. Looking around at these vestiges, I think this old mining town is an acute case study that should make anyone think twice about being complacent about their lot in life. Hundreds of communities today could similarly end up as historical footnotes if some tech-savvy alchemist were to figure out how to power a car with Captain James T. Kirk's dilithium crystals instead of John D. Rockefeller's black gold.

A High Price to Pay

I step through the rectangular void where the building's door used to be, hoisting my leg over the crumbling footings. To be sure, Bankhead is an extreme case of desuetude catalyzed by complacency amidst a poor market for coal. But outside the building, I remind myself that jumping to hasty conclusions is as easy as grinning for a selfie these days. I tend to avoid both.

Posing beside rubble and saying "Bankhead went bankrupt, therefore so will all resource towns" is tempting, but the perplexing thing is that the world didn't stop using coal in 1922. Far from it. Today, 7 billion people gorge on five times as much of the black stuff as a hundred years ago. In fact, Bankhead is not a case of a primary energy source becoming obsolete.

I've read a lot about Bankhead, but I still look to the rubble for practical answers to why Mine No. 80 and the town around it went bust. My attention turns to the historical plaque on the outside wall. Apparently the building I was just in was the miners' lamp house, the gathering place for a hard day of back-breaking, thankless work. Closing my eyes, I time-travel a hundred years in the past, to an early April morning in Bankhead.

At 6:00 a.m., a siren blows, waking the entire community. By 7:45, men are heading to the lamp house for their 8:00 shift. They line up in front of where I'm standing, maybe cracking jokes, shoving each other playfully and cussing about losing at last night's poker game.

Amid the banter, Frank Yakubiec and Mike Perrotti wait to pick up a clean, fully fueled brass safety lamp before they go into the mineshaft. When their turn comes, Frank and Mike get these lamps in exchange for their personal tags, each inscribed with a number.

The two know what this daily tag-for-lamp trade means. They're swapping the valley's natural glories for eight hours of hell inside the bowels of this mountain, bashing and hauling out coal, risking their lives in the process. But they do it, day in, day out — to put food on their families' tables, clothes on their backs.

Frank makes the swap like he does every day. Except on this day — April 22, 1920 — he doesn't make it out.

Records from the Old Calgary Court House diarize that Mike Perrotti "heard coal running down the chute, and about the same instant heard someone shout from inside the chute down below him." That someone was Frank, who "shouted for him to stop the coal from running" and "ordered Mike Perrotti to get an axe and cut a hole in chute . . ."

It was too late. At 10:45 a.m., Frank Yakubiec died under a crush of coal.

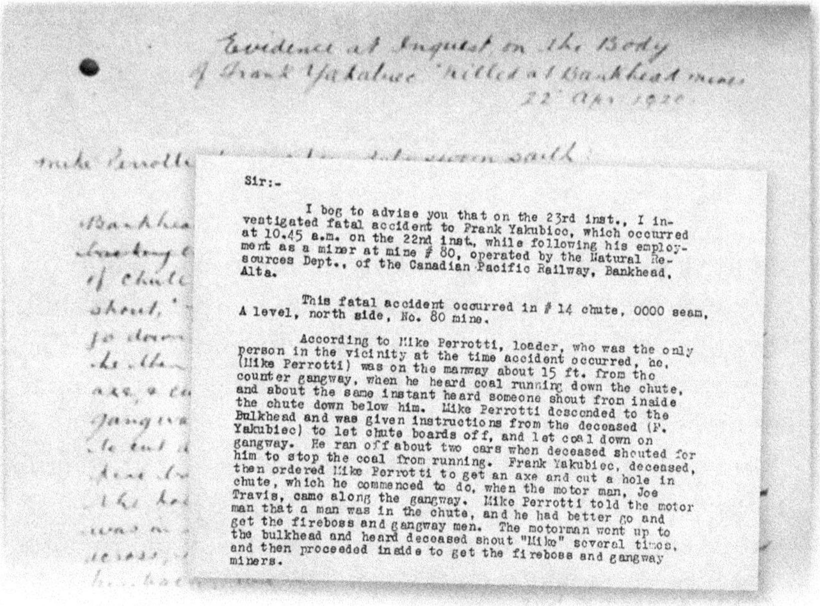

Sir:-

I bog to advise you that on the 23rd inst., I investigated fatal accident to Frank Yakubicc, which occurred at 10.45 a.m. on the 22nd inst. while following his employment as a miner at mine # 80, operated by the Natural Resources Dept., of the Canadian Pacific Railway, Bankhead, Alta.

This fatal accident occurred in # 14 chute, 0000 seam, A level, north side, No. 80 mine.

According to Mike Perrotti, loader, who was the only person in the vicinity at the time accident occurred, he, (Mike Perrotti) was on the manway about 15 ft. from the counter gangway, when he heard coal running down the chute, and about the same instant heard someone shout from inside the chute down below him. Mike Perrotti descended to the Bulkhead and was given instructions from the deceased (F. Yakubiec) to let chute boards off, and let coal down on gangway. He ran off about two cars when deceased shouted for him to stop the coal from running. Frank Yakubiec, deceased, then ordered Mike Perrotti to get an axe and cut a hole in chute, which he commenced to do, when the motor man, Joe Travis, came along the gangway. Mike Perrotti told the motor man that a man was in the chute, and he had better go and get the fireboss and gangway men. The motorman went up to the bulkhead and heard deceased shout "Mike" several times, and then proceeded inside to get the fireboss and gangway miners.

Working in a coal mine in the early 20th century was risky — processes were dangerous and safety standards lax. Accidental deaths like Frank's were common.

Feast or Famine

I open my eyes to take in the precious daylight. "Grim" is the first word that comes to mind.

Standing at the lamp house and feeling its haunting ambience, I know there's no way I could have mustered the nerve to trade a piece of ID for a brass lamp and the possibility of a one-way trip into a clammy black hole.

I reflect on Frank's contribution to providing fuel that heated homes in the thick of winter or stoked steam engines for the pleasure of wealthy tourists riding a train through the Rocky

Mountains. Frank dug out a lot of coal so others could enjoy those things — and gave his life in that pursuit. Yet I'm pretty sure that anyone who enjoyed the benefits of coal from Mine No. 80 had no idea where their fuel came from, nor what sacrifices — including Frank's and many other miners' lives — were made under that mile of mountain.

I don't think men like Frank and Mike were looking for people to stop them on Main Street and thank them for risking their lives so others could have a warm bath. Just the opposite.

From what I've read, Rocky Mountain coal miners were especially stoic. Many were hardworking immigrants from Eastern Europe looking for a better life in Canada. I doubt they found an easier existence here, but I know they wanted to be paid fairly and treated well for the sacrifices they made — compensation and dignity make up for a lot of hardship.

Bankhead appears to have been one of the better mines in the area for working conditions, but, by today's Western standards, it was hardly acceptable. Tension between the United Mine Workers of America and the tweed-suited owners of the Pacific Coal Company led to many strikes over the mine's 20 years of operation.

Indeed, labor strife was a major factor leading to the town's demise. The miners wanted better pay and working conditions, while the owners stereotypically remained intransigent with their profits (or lack thereof).

Coal was a feast or famine business, increasingly the latter, because the cost of production in Alberta was too high relative to the volatile ups and downs of the commodity price. The downturns in the coal market were hard on company coffers and even harder on workers' wallets.

The final, irreconcilable strike began on April 1, 1922. For 76 days, workers refused to gather at the lamp house. Looking around these ruins, I can imagine union leaders and management reps negotiating for hours in a dingy room hazy with cigarette smoke. It must have been hard enough cutting through all that smoke and bullshit in the air, let alone cutting a pay deal.

So there was no deal.

On the evening of June 15, Bankhead began its abrupt decline from bustling, proud mining town to a historical curiosity for hikers.

Powered by compressed air, the dinky was much stronger than a horse and a lot safer than a locomotive making sparks in a gas-filled mine.

An Assault on All Fronts

A gust of wind rustles the leaves, breaking my train of thought and directing my attention to a path. Intrigued, I head over. I arrive at what the miners called a dinky, a hefty air-powered iron cylinder used to haul materials into and out of the mine. It looks like a prop from a steampunk show.

Not far from the dinky, leftover slags provide evidence of what's been written about Bankhead's mine: the quality of its coal was not that good. Sure, the stuff looks black, but it's full of rocks and too wet in composition, prone to crumbling into small grains. Fireplaces like to be fed dry, chunky coal, so, after the dinky hauled coal from the mine, workers had to pick out the rocks, which added a lot of manual labor.

As Mine No. 80 developed, its coal output became finer and finer, which meant the yield of desirable coal chunks was poor relative to the effort required. Bankhead suffered from the most serious of business maladies: poor-quality product and a high cost of production.

I start walking back down the path, my mind wandering farther than my feet.

The operation had challenges on many fronts: labor discontent, poor product and thin profits. Another competitive drag was that Bankhead was a thousand kilometers from the lucrative markets to the east, so transportation costs were prohibitively high.

Bankhead had big-picture business issues, but I also know that local circumstances played a part in taking the place down. The Pacific Coal Company was a sophisticated operation backed by the ample wallet of its parent, the Canadian Pacific Railway. Having corporate backbone was good, but in Alberta, there were plenty of "mom and pop" coal-prone sites. Collectively the little guys could produce fairly large quantities of coal without a lot of up-front money. I've been to some of those sites too, where a few picks, shovels, a donkey, some rickety carts and a cadre of small-time entrepreneurs were enough to get into the coal business.

Whenever the price of coal went up — as it usually did in the dead of winter — the little guys flooded the local market with too much of the stuff. Prices fell quickly in the spring and nobody, big or small, could make a living.

The historical record speaks to brutal competitive circumstances. Owners, managers and workers at Bankhead were constantly challenged to make a buck in this recurrently adverse micro-environment.

After World War I, change hurtled at the coal business like a freight train — literally. Diesel engines powered by oil succeeded coal-fired locomotives in North America. Lessened demand to fill a tender car with coal at pit stops like Bankhead further eroded the mine's sales. The contemporary parallel to this development is obvious: if everyone starts buying plug-in electric vehicles, thousands of gas station owners will be subject to a similar fate.

Hope Ain't No Strategy

I emerge from the woods at the final stop on my tour: what's left of the church on the hill.

There's not much here, just crumbling foundations wrapping around what used to be the basement, now overgrown with shrubs and trees. The central feature is a wide stone

After World War I, change hurtled at the coal business **like a freight train — literally.**

The affectionately dubbed "Stairway to Heaven" still stands strong. The church it led up to, however, is long gone.

staircase that's still in pretty good shape. Affectionately named the Stairway to Heaven, it rises about three meters above the foundations before ending with a sharp drop to the basement. There's no building awaiting, just a view of the heavens.

From the top of the stairs, I survey the ruins and imagine miners and their wives praying for renewed prosperity during those tough times. This place is a case study of life in so many dimensions.

During my MBA, case studies about bankruptcies were coldly analyzed in a sterile classroom. Yet an Excel spreadsheet and a stuffy lecture from a prof in a wrinkled shirt can't begin to capture the human element of business failure and community demise. This old church — where unemployed workers desperately prayed for more food on their table — gives a "distressed" corporate balance sheet entirely different meaning.

The first thing I noticed on my trip to Bankhead was a love story. Despite the hardship, I can sense a love of community nestled here in the beautiful Rocky Mountains. I leave saddened that this love affair lasted only 20 years.

Heading to my car, it occurs to me that my day job may not involve religion, but it is about analyzing business, which is a church of sorts: corporate leaders and their faithful flocks of investors revere making money. When that's not happening, there's much gnashing of teeth, wringing of hands.

I think about the presentations and lectures I've given to people in the energy business over the years. I've observed that, in tough times, especially when oil and gas prices are depressed, the first companies to suffer from competitive pressure are those not at the top of their game. In every case, that underperformance leads management and employees to a strategy of hope and prayer, rather than innovation and improvement. But business, unlike church, is an unforgiving environment, one where success, and failure, are indifferent to such folly.

I always see heads in the audience nod sagely at my commonsense conclusion: if you want to survive in the energy business for the long haul, you must have quality assets; a productive, innovative and disciplined organization that keeps costs low; and access to diverse and lucrative markets that aren't plagued by the ups and downs of seasons and other cyclical factors. I realize that isn't a message everyone wants to hear though, especially those in financial trouble.

Instead, maybe I should encourage my audiences to forget their spreadsheets, stop listening to me and plan a trip to Bankhead. Anyone who comes here will realize that merely hoping for business success can lead to their own Stairway to Hell — a place where too many have found their end.

QUESTIONS AND ANSWERS

Introduction

At its core, "Stairway to Hell" is about the travails of a company under competitive siege, additionally plagued with bad assets and poor labor relations. But there are other layers of complexity, because Bankhead, the associated village, was a one-company town.

The questions that follow will help you:

- recognize the early signs of competitive business assault, disruptive change and organizational distress
- understand corporate defense mechanisms to fend off competition and survive disruption
- learn about the gap between those who work to supply energy and those who consume it
- explore the relationships between communities and the industries that support the local economy
- reflect on your own organization's circumstance and its ability to respond to disruptive business forces

Often, large-scale business failure is blamed on external forces like "the economy" or "the government." After discussing "Stairway to Hell," you'll see that culpability for disruption is not so easy to determine. But you will be well equipped to recognize vulnerability to change and suggest preventative measures.

Questions

1 Red flags of impending financial trouble are remarkably visible in challenged industries and companies. How management teams choose to address the warning signs, or not, determines corporate success, or demise.

A What were the internal operational warning signs that the Pacific Coal Company was going to close its mine at Bankhead?

B What were the external economic and technological warning signs of Bankhead's looming demise?

C Are any of these warning signs affecting your organization? If yes, what are they?

2 Many industries are undergoing disruptive changes right now, much as coal mining at Bankhead was a hundred years ago. For example, lots of retail, media and energy companies are struggling. Often it seems that the deeper the distress during periods of change, the greater the loss of objectivity among decision makers.

A Discuss the concept of "relying on hope" as a strategy, especially as a response during periods of extreme competitive pressure.

B What is the opposite of relying on hope?

C When overwhelmed by change and disruption, management's judgment can be clouded. How would you characterize your organization's response to the forces of change? To what extent is it a strategy of hope?

3 Industrial disruption is happening in every industry. For example, electric cars are conjectured to take away significant market share from combustion engine vehicles and potentially lessen the demand for oil.

- A In the context of oil, discuss the characteristics of companies that may follow the same fate as the Pacific Coal Company. What are the traits required for survival through disruption?
- B Imagine you are an investor in oil companies. Taking the lessons of Mine No. 80 into account, what investment strategy would you recommend?
- C Now imagine you are an investor in renewable energy companies. Again, taking the lessons of Mine No. 80 into account, what investment strategy would you recommend?
- D Do you believe that, with the growing popularity of electric cars, all oil companies are doomed? Why or why not?

4 "Stairway to Hell" has a weighty social overtone — the demise of an entire town.

- A What do you think would happen if your local newsfeed reported that everyone in a nearby town of 1,500 was going to lose their jobs and have to relocate?
- B Who or what would receive the bulk of the blame for such an event? Who *should* receive the blame?
- C Quite often, towns, cities and even entire countries grow around a nascent industry. Discuss some risks and benefits of excessive reliance on one industry, or even one company, in a locale.

5 On the death of Frank Yakubiec, I conjecture, “anyone who enjoyed the benefits of coal from Mine No. 80 had no idea where their fuel came from, nor what sacrifices — including Frank’s and many other miners’ lives — were made under that mile of mountain.”

- A Discuss the societal disconnect between those who work to supply energy and those who consume the benefits. What are the potential consequences?
- B Do you know where your energy comes from, or what safety risks are taken by those who work to bring you the comforts you enjoy?

6 Poor and contentious labor relations were contributing factors to Mine No. 80’s demise.

- A Discuss the culture and labor relations at the Pacific Coal Company.
- B Discuss the culture in your organization. Does it foster innovation and competitiveness?

7 “Stairway to Hell” takes place a hundred years ago and focuses on the dynamics between the Pacific Coal Company’s management and workers. Today, government and business are more intertwined. As you answer the following questions, think about the story within the complexity of today’s socio-regulatory environment.

- A What do companies or their industry associations typically do when confronted with competitive assault, disruption and potential demise?

B Should governments intervene to save jobs in communities subject to potential demise?

C What can politicians and policy makers do to avoid a Bankhead-type situation?

8 The technology threat of diesel engines was another contributing factor to the Pacific Coal Company's demise.

A What do you imagine the CEO of the Pacific Coal Company thought upon seeing the first diesel engine pull into Banff without requiring a fill-up of coal?

B Thinking about your own business or energy circumstance, what do you see as early indicators of potentially disruptive technologies?

C Are the decision makers in your organization paying attention to leading indicators of technological change?

Answers

1 A **What were the internal operational warning signs that the Pacific Coal Company was going to close its mine at Bankhead?**

The story states that the company "suffered from the most serious of business maladies: poor-quality product and high cost of production." Yet this ailment was not realized until much later in the mine's 20-year history. During the mine's early life, the coal's quality was actually quite good, which is why the company expended so much capital developing the resource to start.

A big early warning sign was the deterioration of coal quality, especially the propensity of the chunks to crumble. From an accounting perspective, the financial statements would have shown rising operating costs. A bit more investigation would have revealed the cause of those rising costs to be extra labor (to sift out rocks and chunks) per unit of coal produced. Asset deterioration in resource businesses like coal, oil and gas often happens slowly.

The Pacific Coal Company was in a cyclical business with significant commodity price volatility — poor operating performance was frequently buried under, or blamed on, the low profitability that typically accompanies weak product prices.

Finally, poor labor relations should have been a warning sign. In the early 20th century, strained relationships between workers and management were the norm, with workers often at the mercy of their masters, so labor action at mines like Bankhead was probably diarized as the cost of doing business.

Yet strikes and other labor action are always costly. Today, the labor antagonism and dismal working conditions at Bankhead wouldn't be tolerated. However, the broader lesson is that worker dissatisfaction is never good for productivity, hence profitability and investment returns.

Over many years in the investment business, I've found that the best way to gauge employee satisfaction in a portfolio company is to talk to workers below the management level. Taking someone for lunch or a drink after work is a good way to get the scuttlebutt. In one instance, I was made aware of abusive behavior by a senior management member. After further enquiries and an internal investigation validated the employee's assertions, the manager was let go.

1 B What were the external economic and technological warning signs of Bankhead's looming demise?

Clearly the volatility of coal prices was concerning. The company closed its doors during a strike, probably blaming labor issues. But the reality was that the assets were poor and couldn't sustain the big downdrafts in coal prices that accompanied the seasons.

One red flag was the low barrier to entry in the region. When prices rose in winter, or due to other positive macroeconomic factors, competitors from coal-producing regions elsewhere in Alberta could quickly ramp up their output, creating a regional glut that soon depressed prices again.

The demand side of the business should have been a concern, too. Alberta didn't have enough of a regional market to absorb all potential output. Market access — the ability to tap into the more lucrative central and eastern North American customer base — was hampered by high rail transportation costs. In short, from afar, Bankhead coal could not compete with the superior, cheaper and more accessible product from regions like Pennsylvania.

Finally, the potential for substitution was a warning sign. In the early 20th century, diesel locomotives were starting to replace coal engines. Owned by the Canadian Pacific Railway (CPR), Bankhead was located on a major east-west rail line in western Canada, with nearby Banff serving as an important refueling stop for coal locomotives. Surely CPR management would have been aware of this transition to diesel fuel, away from Bankhead's coal. In fact, I believe this knowledge would have contributed to management's decision to abandon the mine.

Cyclicality can be a company's best friend during commodity price-up markets. But it's a nightmare when prices weaken, especially for companies that have product quality and operational issues.

1 C Are any of these warning signs affecting your organization? If yes, what are they?

Warning signs are relentless, so I'd be skeptical if anyone answered "no." If you're having trouble identifying specific warning signs, consider what companies offer competitive products or services. Are they bringing new products to market that may be better or lower-cost than your company's? Internally, consider markers like trends in quarterly profitability — waning margins are often signs of trouble. These are only two of many possible examples. What do you see affecting your organization?

2 A Discuss the concept of "relying on hope" as a strategy, especially as a response during periods of extreme competitive pressure.

Learning about Bankhead, I have this strong sense that both management and workers were hoping for better days, hoping for prices to go up, hoping for the resource quality to improve, and so on. This is neither unique to Bankhead, nor to the era. I've seen this behavior many times in my career analyzing companies.

In the business world, "hope" is usually couched in the financial vernacular of "expectation." I'm always cautious when company leaders confidently state "We expect prices will go up" or "We expect the economic conditions for our business will improve." Or the ultimate strategy of hope: "We expect a new government will be elected to help."

Expectations are most often linked to macroeconomic, political or other factors out of a company's control. Managing exogenous risk factors is, of course, part of a corporate leader's responsibility. Relying too much on beliefs that external influences will sway in a company's favor is a risky strategy of hope.

2 B What is the opposite of relying on hope?

Identifying and acting on what a company can control and improve is the opposite of hope-as-a-strategy. The best way to avoid demise is to stay vigilant about being competitive (in other words, proactively innovative), while being operationally disciplined and strategically positioned to fend off market assault.

For instance, a company may not be able to control the market price for its product, but it can figure out how to reduce costs to remain competitive. That's an example of taking control. Admittedly, in Bankhead's case, operating in such a volatile price environment was difficult (as it still is for many seasonal commodity businesses).

Adapting to macroeconomic, technological and political forces of change is one of the most challenging management tasks. Too often, in the midst of accelerating change and competition, denial of the seriousness of those forces can lead to paralysis and a retrenchment into a strategy of hope.

Proactive business strategies should focus on flexibility and adaptability, with emphasis on tasks in a company's control — for example, optimizing operations and reducing costs — where expectation can be met with accountability.

2 C How would you characterize your organization's response to the forces of change? To what extent is it a strategy of hope?

In answering this, consider writing down the narratives and directives your organization offers to its employees and investors. How many of them are strategies of hope versus strategies in its control?

FACILITATOR'S TIP

Question 2c may also be approached from an industry, community or civic perspective.

---

3 A Discuss the characteristics of companies that may follow the same fate as the Pacific Coal Company. What are the traits required for survival through disruption?

In the example given, the sales potential of electric vehicles is a threat to the transportation segment of the oil business. The erosion of oil consumption has the potential to create a surplus of petroleum products like gasoline, thereby weakening oil prices. All else being equal, the profitability of oil producers will fall as a result. So, the first and most detrimental characteristic of companies that are vulnerable to disruption is a high cost of production.

Conversely, enduring companies are efficient and doggedly determined to continually improve operations and reduce costs. For instance, while Mine No. 80 went into demise, other, lower-cost Alberta mines endured. The Atlas Mine by Drumheller was among several that operated six decades longer, into the 1980s. The main differences were better product quality and operations, which enabled Atlas and other survivors to withstand down cycles in price.

In the case of Bankhead, only regional demand was affected by the substitution of coal locomotives with diesel. Globally, coal consumption continued to rise for more than a century, and plenty of investments yielded lucrative returns in places where coal continued to be copiously consumed. Just not for the Pacific Coal Company.

So, another characteristic of companies vulnerable to disruption is being trapped in a local economy with no ability to tap into growth markets elsewhere. Again, the converse to this question highlights a characteristic of an enduring company amidst potential product substitution: the flexibility to access alternative markets.

The effect of price erosion and lack of market access on high-cost resource producers will always be the same: ongoing financial stress and, in the worst case, distress. In other words, the potential for a one-way trip down the Stairway to Hell.

3 B Imagine you are an investor in oil companies. Taking the lessons of Mine No. 80 into account, what investment strategy would you recommend?

In theory, the principles for commonsense investing are not complicated: choose companies that have high-quality resources (in this case, oil-bearing rocks) paired with a culture of passionate, productive employees who know how to continually innovate to bring down costs.

Proverbially "lean and mean" organizations are able to survive cyclical downturns and come out stronger on price recovery. Inefficient, weaker peers with poor assets and employee culture (like the Pacific Coal Company) are among the first to experience demise. This Darwinian dynamic leaves the fittest companies in a competitive fray with greater profit potential.

Eras of energy transition for resource producers are different from normal cyclical ups-and-downs. During a transition, the demand for commodities begins to wane permanently, much as the consumption of coal started to retreat in the Banff area as diesel fuel took over in locomotives.

Yet, in the world of primary energy sources — such as wood, coal, oil or natural gas — regional transitions can take years, if not decades. There's always residual demand for a commodity. After all, many people still burn wood. The question then becomes, Which company or companies will supply the residual demand?

Clearly, the lowest-cost producer with the best product and delivery service will endure. Again, all wood suppliers didn't disappear when we transitioned from wood to coal in the 19th century, but the high-cost ones did.

So, the strategy for resource investing is pretty obvious: stick with the lowest-cost, most efficient producers with the best assets and operations. We also learned from the previous question that companies with access to markets beyond the local area have an advantage. While not essential, such geographic flexibility further reduces a company's risk of obsolescence.

3 C Now imagine you are an investor in renewable energy companies. Again, taking the lessons of Mine No. 80 into account, what investment strategy would you recommend?

At first blush, energy systems based on renewables, like wind turbines and solar panels, may seem immune from the disruptive circumstances that plagued the Pacific Coal Company. But they are not.

In highly competitive industries, there is no immunity from technological change or bad management. Who knows what new energy technology may move into a region, threatening the status quo? Ill-equipped management or bad culture can't deal with disruption. Just like any other industry, the renewable energy business is littered with bankruptcies.

Nor are renewables immune from locational considerations. Solar panels and wind turbines are often suppliers to highly localized markets, making them vulnerable to any macroeconomic risks afflicting the communities they serve.

As in any industry, old or new, there are low-cost and high-cost manufacturers, good operators and bad, well-positioned companies and those that are poorly situated. The type of business doesn't matter (fossil fuels or renewables). The lessons of Bankhead and the Pacific Coal Company are universal.

3 D Do you believe that, with the growing popularity of electric cars, all oil companies are doomed? Why or why not?

I ask this question because many people think so, and I've often debated against the conjecture that "all" oil companies will die out over the next few decades.

It's easy to understand why people think they will. The dynamic of electric vehicles (EVs) displacing gasoline engines today is superficially analogous to diesel locomotives displacing coal-burning steam engines in the early 20th century.

As we learn in "Stairway to Hell," Bankhead was doomed in part because the minehead was next door to the fueling stop at the popular tourist resort of Banff. Today, this is akin to EVs bypassing gas stations in favor of electric-charging infrastructure.

Yet think about this question with respect to what you learned from questions 3a to 3c. The Pacific Coal Company went out of business, but were all coal companies doomed because of the entry of diesel locomotives? The answer is clearly no.

Over time, the demand for oil will be reduced at the margin due to EV market penetration, and so the price of oil is likely to weaken. However, as the discussion of prior questions has demonstrated, the most vulnerable companies are those that have a high cost structure, lousy assets and poor operations. Strong companies with converse characteristics will always survive.

Oil, like coal, has many uses other than fueling transportation. Only 40% of oil consumption goes to moving light-duty vehicles. Much of the rest goes to industrial products like lubricants, plastics and petrochemicals.

And consider this: coal-fired steam locomotives are no longer around, but coal is still used copiously in market segments other than train transport. By analogy, petroleum-fired cars are likely to cede to EVs, but oil and the companies that produce hydrocarbons will be around for much, much longer.

4 A **What do you think would happen if your local newsfeed reported that everyone in a nearby town of 1,500 was going to lose their jobs and have to relocate?**

I know this is a bit rhetorical, because the answer is obvious. There would be considerable concern about the fate of the workers and the town's residents, fueled by media coverage. We see this when a local auto plant or other major manufacturing facility shuts down. Workers come out with placards, and the hardships are amplified by social media. Such outcries elicit sympathy and potentially action to try to arrest the change.

In the early 1920s, 1,500 people was a lot in proportion to the general population — as a percentage it was certainly a lot more than today. I haven't scoured newspapers of the time to see how the announcement of the shutdown was received, but I imagine it would have devastated the townspeople. Many coal workers found replacement jobs in nearby oilfields like Turner Valley, so the impact of the shutdown may have been muted. Regardless, families were displaced, which is never easy.

Today we receive news (or what's construed as news) instantly, and we know bad news travels faster than good. Nothing is muted in our newsfeeds, especially armchair opinions and conjectures. Consequently, I believe social media often pushes a force of *resistance* rather than change. A town of 1,500 people shutting down today would likely receive greater outcry than in the past, leading to more external support to find the ways and means to preserve jobs and community.

FACILITATOR'S TIP

In question 4a, two points should come out:

- Industrial disruption has societal consequences that are easily publicized.
- Public outcry can act as a force of resistance against change.

4 B Who or what would receive the bulk of the blame for such an event? Who *should* receive the blame?

Everyone will have a different opinion. In my assessment of how society responds to disruption, the people who are most directly affected typically blame the company (which means its leadership) or the government.

The real force of change for disruption often comes from the microeconomics of business: stronger, disruptive competitors. In other words, better companies that have product, process and market advantages. That's what worked against the Pacific Coal Company.

However, from what we've learned in question 3, the culpability ultimately lies with company leadership that doesn't have "the right stuff" to fend off new entrants coming into their market. I believe that's an appropriate assessment. Successfully navigating the risks of competition and economic malaise is what good management is paid to do.

4 C Discuss some risks and benefits of excessive reliance on one industry, or even one company, in a locale.

Again, this is rhetorical, but sometimes the most obvious questions represent the elephant in the room. One-industry, or one-company, communities have the burden of what is called "concentration risk" — their fortunes are heavily dependent on the performance and competitiveness of the concentrated economic activity.

The most obvious benefit of industrial concentration is community prosperity when times are good. Yet when disruption knocks on the door, that risk brings disproportionate misery into town. That's why leaders of industrially concentrated communities often argue in favor of economic diversification during the good times.

5 A Discuss the societal disconnect between those who work to supply energy and those who consume the benefits. What are the potential consequences?

Energy systems have two bookends: suppliers of raw energy commodities (like coal and oil) and users of the amenities they ultimately provide (heat, light, mechanical work). Then there are all the processes between. It's like the gap between ranchers/farmers and restaurant diners. The latter often have no idea where their food comes from and who toils behind the scenes to make it happen. Ditto for energy.

Supplier-consumer disconnects — being blind to the realities of what happens behind the scenes — can lead to multiple divisive issues, including but not limited to:

- devaluing the importance of upstream energy workers, leading to apathy about their plight in the face of transitional disruption
- being complacent about energy security
- misunderstanding where our day-to-day amenities originate, potentially leading to poor energy policies

Other Energyphile stories address this disconnect between energy producers and consumers. If you're interested in reading more, check out "Alfred Dickie's Utility Bill" and "Nobody Tips a Scandiscope."

5 B Do you know where your energy comes from, or what safety risks are taken by those who work to bring you the comforts you enjoy?

I hope the death of Frank Yakubiec got you thinking about what goes on behind the scenes in the pursuit of societal amenities. As you consider this issue, think about each major energy amenity: heat, light, electrical power for various appliances and mechanical work (like turning wheels).

For heat, your source of supply may be electricity or natural gas. Do you know where they're sourced from? Even if they originate from Canada, Europe or the United States, where safety regulations are strict, job conditions can still be dangerous. Inclement weather can further raise the risk of industrial accidents. Think about workers repairing high-voltage power lines, drilling for oil on deepwater rigs or climbing up high on wind turbines to service them.

But not all energy is sourced from within North America. Do you know where your gasoline or diesel originates? It's possible that it's from barrels of oil imported from authoritarian countries with lax safety and human rights standards. New-age batteries that power your cell phone or electric vehicle may contain cobalt, mined by child labor in an African artisanal mine.

This issue is tackled extensively in another Energyphile story, "Nobody Tips a Scandiscope."

6 A **Discuss the culture and labor relations at the Pacific Coal Company.**

What's important in discussing this question is recognizing the mechanism that corporate leadership used to weather down cycles and competitive assaults. The first point of leverage for the Pacific Coal Company was reducing the workforce or asking for things like wage concessions to cut costs. Of course, this created a toxic environment that amplified management-worker polarization. Poor labor relations, in addition to uncompetitive assets, was a heavy contributor to the mine's poor productivity and ultimately its demise.

Today, stricter government labor regulations and safety standards have improved worker conditions. However, bad culture due to a panoply of management-worker tensions can still reduce organizational productivity, hence competitiveness.

FACILITATOR'S TIP

Explore the converse: how positive labor relations and a culture of innovation is a more productive counter to competitive assaults.

---

6 B Discuss the culture in your organization. Does it foster innovation and competitiveness?

In my experience, this is a sensitive topic, yet one of the most important. The objective is to flush out hard realities about your workplace culture, specifically worker-management relations.

As you discuss your organization's situation, refer to 6a. When faced with competitive forces, does your management tend to fall back on that same mechanism? Or does it foster a culture conducive to winning by empowering employees to constantly innovate?

7 A What do companies or their industry associations typically do when confronted with competitive assault, disruption and potential demise?

Denial of change is usually the first response, followed by blaming others. Disruption is often accompanied or facilitated by government policy (a whole separate story!), so pointing fingers at the government or other macroeconomic circumstances is also quite typical.

When competitive business disruption translates into financial underperformance — or outright distress — the next phase is to clamp down on the workforce, which is the subject of question 6.

By this time, it's often too late for management to recover the business. Repositioning the company to compete generally requires more capital, but the financial stress typically means there's no money available. Investors are loath to back a weak competitor with fresh capital, so, notwithstanding macroeconomic reprieve, the company is likely doomed.

7 B Should governments intervene to save jobs in communities subject to potential demise?

The answer depends on your political stripe. Left-leaning individuals will argue for intervention, while free-marketeers will call for a hands-off approach. When an entire community is under siege, the latter is a tough call.

Often governments will step in to save a significant number of jobs, especially if a whole town is at stake. (It's not facetious to acknowledge the political reality: jobs equal votes.)

Government support, financial or otherwise, should be based on an understanding of the root causes of downturn and potential demise. There's no point propping up a company (or companies) that will never have "the right stuff" to compete, especially if the assault on its competitiveness is expected to be inevitably fatal.

Where problems arise — and difficult decisions must be made — is when there are deep structural changes in an industry that affects localities overly dependent on uncompetitive businesses. That was evident in "Stairway to Hell" — it's not likely that any amount of government assistance could have led the company back to health.

Cyclical downturns with an expectation of recovery can warrant interim support. An extreme example of this was during the 2008 financial crisis, when government bailed out North American automakers to avoid bankruptcies and mass unemployment. The industry recovered, and government got its money back.

Often, cyclical downturns can be healthy because they weed out uncompetitive businesses and toughen up those that weather the tough times. On recovery, the strong companies emerge stronger and more efficient. In free markets, it's very much survival of the fittest.

7 C What can politicians and policy makers do to avoid a Bankhead-type situation?

In locales that are heavily dependent on one industry or company, the business *is* the economy and as such is codependent with government. In these cases, the best government policies are those that work proactively and collaboratively with codependent businesses to ensure competitiveness, before it's too late.

To keep citizens prepared for change, government incentives for innovation and skilled-workforce training should be key areas of focus. Fiscal policies can provide tax incentives for innovative behavior.

Ultimately, diversification is the best policy, and part of a government's role is to attract and foster different types of businesses to lessen concentration risk.

8 A What do you imagine the CEO of the Pacific Coal Company thought upon seeing the first diesel engine pull into Banff without requiring a fill-up of coal?

If what I've seen in my career is any indication, the CEO is likely to have dismissed the event as mere novelty. Denial of change is the first response of leaders who are entrenched in a business paradigm. Yet this is not to say the CEO is dumb. After all, denial is a foible many of us suffer from.

The other side of the coin is that it's also folly to have a knee-jerk reaction. I'm the first to agree that any capable CEO shouldn't abruptly pivot their business with one data point of experience.

So when is the right time to take competitive challenges — technological or otherwise — seriously enough to warrant major organizational change? It's not easy to know, and every situation is different. But being paranoid about the deleterious effects of competition, macroeconomic downturns and disruptive change is the job of a capable C-suite.

FACILITATOR'S TIP

Have the group discuss when management *should* start taking a technological threat seriously.

---

8 B Thinking about your own business or energy circumstance, what do you see as early indicators of potentially disruptive technologies?

While every business is different, none is immune from disruptive technology.

In this highly dynamic period of energy transition, things can change quickly. For example, we know that solar and wind energy are a threat to the entrenched fossil fuel industries. But advances in small-scale nuclear and even fusion technologies could pose a challenge to renewables. And what if fossil fuels were to come up with technology that reduced CO_2 emissions to zero? No segment is immune. Everything is dynamic.

In the energy business, threats can emerge from any part of the system. For example, at the consuming end, electric vehicles pose a threat to manufacturers of internal combustion engine cars. Upstream, the threat is a challenge to oil producers. So not only do leaders of oil-producing companies have to monitor technological challenges from within their industry (innovations introduced by other oil companies), but also from far down the value chain (from the automotive business). It's not easy staying on top of everything.

As you list the potentially disruptive technologies, classify them first from within your sphere of business (your competitive peers), then from outside your sphere (challenges downstream or peripherally).

8 c Are the decision makers in your organization paying attention to leading indicators of technological change?

When you answer this, you may be quick to chastise management for not being fast enough to recognize impending change. And it may be true. But paying attention to leading indicators is only the start of a long competitive battle.

Large companies have chief technology officers whose role includes monitoring the competitive landscape for threats and opportunities. Identifying threats is the easy part. What's much harder is convincing others, like the CEO, of the need to pivot amidst disruption. Then there's determining when and how to react — deciding on a response strategy and leading change is the *really* hard part.

In reality, it's not up to one or two people. Successful companies are culturally programmed to constantly be aware of potential threats, assess them and react accordingly. None of that is possible without first having the antennae out for sensing the looming changes. After that, acknowledging those leading indicators is key to overcoming denial and responding.

FACILITATOR'S TIP

If anybody feels leadership isn't paying attention, discuss what they'd do if they were in charge.

FACILITATOR'S GUIDE

Come Together, Move Forward

Whether it's a corporate planning session, a class discussion or a social book club, an Energyphile discussion encourages critical thinking, sparks lively conversations and helps build a community equipped with tools needed to make thoughtful, well-informed decisions for a better energy future.

The questions in this discussion guide invite participants to dig deeper into the issues explored in the story and gain a broader understanding of the forces of change that affect our energy circumstance. Encouraging people to learn from the past, draw parallels to the present, then apply to the future, these questions embody the Energyphile philosophy.

As the facilitator, you will ensure the conversation stays on track, its objectives are met and everybody walks away with a better grip on how they can move forward, while having had stimulating conversation — and some fun — along the way. This guide will help you plan and prepare for a fruitful discussion.

Planning

WHAT'S YOUR OBJECTIVE?

Start by defining your goals. Every gathering of people is different. Will this workshop be used as part of a team-building program? Is your organization facing a particular business issue you'd like your C-suite to tackle in a strategy session? Or do you just want to provoke thought at a family barbecue? Whatever the reason, bringing people together fosters a greater sense of mutual understanding.

WHO SHOULD YOU INVITE?

The occasion and objectives will determine who to include. If your guest list is less prescribed, consider inviting people from different backgrounds, companies or departments. One of the more interesting discussions we've seen included employees from across a company, both office and field staff. Another robust session saw a group of accomplished friends gather around a dinner table.

HOW MANY PEOPLE?

Aim for 6 to 10 people, preferably not more than a dozen. Any more and you can separate into breakout groups. Smaller groups allow for more questions to be covered with greater depth.

HOW LONG?

A good length is 2 to 3 hours. If you want to offer a longer workshop, consider breaking it up into discrete sections — for instance, cover all policy-oriented questions in one — or incorporating other activities, like having groups research different questions and report back.

WHERE?

Of course, there's always the boardroom, but think about leaving the office. Getting people out of their usual environment and routine turns it into a more social outing and may help shift the dynamics. Many coffee shops and restaurants have private rooms. The library probably rents rooms for free, or there may be innovation hubs that offer space. Check out community centers or coworking spaces for rentals, too.

If you're venturing outside the office, ensure the space can accommodate your tech requirements and other amenities, such as catering. If you're rounding up a geographically scattered group, try Skype, Google Hangouts or Zoom.

Preparing

Use this checklist of tasks and suggested timeline to prepare for the day of discussion.

TASK	LEAD TIME	COMPLETED
Book the venue and any required tech.	2 months	
Invite participants (request RSVP).	6 weeks	
Order hard copies of the discussion guide for participants from Amazon or Indigo.* (If participants are responsible for purchasing their own, ensure they've ordered *at least* two weeks in advance.)	6 weeks to 1 month	
If you plan to record the workshop, take notes or generate action items, delegate someone to be responsible for that.	1 month	
Order catering.	3 weeks	
Distribute discussion guide to participants.	2 weeks	
Email the energyphile.org link to the story so people can read its vignettes and listen to the audio version. (Some will prefer audio over print.)	2 weeks	

* For orders of 10+ copies, Energyphile offers a discount. Your order must be received at least six weeks before your event. Contact hello@energyphile.org.

TASK	LEAD TIME	COMPLETED
Email any other supplementary material participants should read in advance (news stories, internal documents).	2 weeks	
Read the introduction to the questions to understand their general considerations and themes.	1 week	
Read the questions, prioritize them based on your objectives and allotted time. Add any of your own.	1 week	
Organize the tech and tools you'll need (whiteboard, flip chart, markers, sticky notes, laptop, projector, screen).	5 days	
Finalize other material you intend to use (PowerPoint deck, handouts).	4 days	
Send a reminder to participants to read the story and questions (but not the answers!). If you plan to cover select questions, you may want to let participants know which to focus on. Remind them to bring paper/pen or laptop for note-taking if they desire, and of day, time and venue. If you plan to record the session, note that, too, in case anybody has concerns.	3 days	
Refresh yourself with the questions you've identified as your priorities and determine how long to spend on each. Think about how you'd like the discussion to unfold.	1 day	
Make name tags or cards for all participants if they don't know one another.	1 day	

During

Kick things off by stating what you want participants to gain from the discussion: Learn takeaways they can apply to their day-to-day role? Determine a course of action for a specific strategic initiative? Or is it purely meant to get brains working and people talking? Ask the group what they hope to get out of it, too.

Cover the housekeeping considerations: how long the discussion will last, format, when breaks will happen, where washrooms are. If you intend to record, remind people of that.

Ask everybody to briefly introduce themselves: name, where they work/what they do, what they hope to learn.

Establish the ground rules:

- Respect all points of view — maintain an open, supportive forum for every person to express their thoughts and to learn from one another.
- One speaker at a time. Give people space to speak.
- Avoid side conversations or other interruptions.
- There is no right or wrong. Disagreement is welcome but don't make it personal. (Address the *idea*, not the person who shared it.)
- Mute your phones.

As you guide the group, weave the past, present and future throughout the conversation. Remember the Energyphile philosophy: learn from the past, draw parallels to the present, apply to the future.

If action items will result from the discussion, ensure these are documented as you go.

As the discussion progresses, watch the time to ensure you're sticking to the pace you've determined.

Allow about 15 minutes at the end to review the action plan and summarize the discussion's main takeaways with the group. What were their most surprising "Ahas"? Did their views change as a result of the discussion?

TIPS FOR MODERATING

Your role is to keep the discussion on track while creating an environment that ensures all participants feel comfortable to speak their thoughts, even if dissenting. These tips can help you do that.

People may need time to warm up. If that first question has you facing a silent room, try paraphrasing it or suggesting other ways they may come at it. If you know someone has expertise or interest in the area, invite them to share their thoughts if they're comfortable. You could also move on to another question — perhaps one that relies more on subjective takes than deep subject knowledge — and return to the first later.

Guide the conversation to keep participants engaged and focused (very important!). People can easily get sidetracked when exploring contentious questions — make sure they stay on topic.

If the discussion does become difficult or tense, pull it back to the story to re-establish common ground and remind people of the lesson.

Direct the conversation, but don't dominate it. Stay neutral and focused on listening, rather than offering your own opinion.

Make sure everyone has a chance to be heard — and understands the gift of listening. This may mean you have to clear space for people. Moderate frequent contributors if they start to dominate, and stay attuned to the quiet participants. Watch body language. Does somebody seem uncomfortable with a topic or speaker? Does a normally silent person look like they have something to say? Use those cues to steer the conversation.

Encourage people to unpack their responses when appropriate. Prompt them with questions like "How?" "What leads you to that view?" "Can you give me an example?"

Use natural segues or lulls in conversation to move on to the next question. If the discussion shows no signs of winding down, blame the clock for cutting it off and proceeding to the next.

After

Within a week, follow up with notes, the recording, action items and anything else promised.

If you're considering offering such sessions regularly, send participants a survey to gauge what worked, what didn't, areas for improvement or change and preferred next topics or stories.

Finally

Enjoy the discussion!

SOURCES CITED AND IMAGE CREDITS

Sources Cited

Page 3: "More than 85% of the world's insatiable needs..."
BP, *BP Statistical Review of World Energy 2017*
bit.ly/bp_energy

Image Credits

All objects shown are in the collection of Peter Tertzakian, with the following exceptions:

Report to the Old Court House, Calgary, AB, Canada (page 8, foreground) from the archives of Parks Canada.

Mike Perrotti's witness statement (page 8, background) from the Provincial Archives of Alberta.

Photography by Peter Tertzakian: pages 3, 8, 10 and 14.

ABOUT THE AUTHOR

THE QUINTESSENTIAL ENERGYPHILE, Peter Tertzakian has devoted his career to energy, first as a geophysicist, then as an economist and investment executive. He's written two bestsellers — *A Thousand Barrels a Second* and *The End of Energy Obesity* — and is sought around the world as a trusted, engaging speaker. Energyphile is the culmination of his passion and knowledge.

DISCOVER MORE AT

ENERGYPHILE.ORG

- Let your curiosity wander in the interactive museum
- Explore artifacts in the Energyphile collection
- Hear the stories come to life in the audio productions
- Check out new stories and content

Continue the conversation with Energyphile Sessions

If you enjoyed this book, explore others in the Energyphile Sessions series. Covering a range of issues, these discussion guides will get you thinking about the business of energy in a whole new way. More are always being added, so check back regularly.

energyphile.org/sessions

CPSIA information can be obtained
at www.ICGtesting.com
Printed in the USA
LVHW091213090120
643001LV00001B/25/P